HOMEGROWN ENERGY

POWER FOR THE HOME AND HOMESTEAD

FINDER'S GUIDE NUMBER FOUR

HOMEGROWN ENERGY

POWER FOR THE HOME AND HOMESTEAD

Gary Wade

Cover by Chuck Hathaway

OLIVER PRESS
WILLITS, CALIFORNIA

CHARLES SCRIBNER'S SONS
NEW YORK

Library of Congress Card Number 74-84300
ISBN 0-914400-03-7

First Printing December 1974

OLIVER PRESS
1400 Ryan Creek Road
Willits, California 95490

CHARLES SCRIBNER'S SONS
New York

CONTENTS

INTRODUCTION

This guide is an attempt to ease the problems faced by anyone who wants to find out "who" makes "what." Those companies which supply alternative energy sources have been located and identified. Their catalogs have been analyzed and their projects or products have been broken down into useful catagories. If you want to find out who makes a particular energy product, it is a simple matter of consulting the index. Refer to the companies indicated by the index, and the energy products will be listed for you. There is also information in this guide about the companies' catalogs. We have tried to solve the problem of finding out who makes what alternative energy sources by doing much of the preliminary, time consuming work for you.

Only companies with a catalog or acceptable equivalent are included in this guide. No responsibility is assumed for any claims made by the suppliers about their products or services. It is hoped that subsequent versions of this guide will remedy such unintentional omissions and errors as might occur, and that frequent revisions will keep the guide abreast of the world of alternative energy.

Although we would like to list prices, the facts of modern life with its changing prices, shortages, strikes and inflation make it futile to do so. Only the companies can, at any given time, quote an accurate price. The manufacturers or distributors are affected by shortages like everyone else, and this causes delays and changing prices. And even though they try, the U.S. Postal Service is often hampered by large volumes of mail and doesn't keep up all the time. For these reasons, patience may be necessary in dealing with the suppliers during these times.

HOW TO USE
HOMEGROWN ENERGY

Homegrown Energy consists of two main indexes: a MASTER INDEX and a COMPANY INDEX. The MASTER INDEX is a list of products alphabetically arranged. If you are looking for a specific product, first check for it in the MASTER INDEX. The MASTER INDEX will then tell you which company or companies offer the specific product.

MASTER INDEX

SOLAR water heater plans
- Earthmind
- Solar-Pak
- Solar Water Heater Co.
- VITA, (Volunteers in Technical Assistance), Inc.
- Zomeworks Corp.

SOLAR water heaters
- Edmund Scientific Co.
- Fafco, Inc.
- Rayosol
- W.R. Robbins and Son Roofing Co.
- Sol-Therm Corporation
- Solar Energy Co.
- Solar Water Heater Co.
- Sunsource
- Sunwater Company
- Zomeworks Corp.

Next check the COMPANY INDEX to find out more about each company listed. The COMPANY INDEX offers information about each company, the range of its products, as well as details about its catalog. If the description of a company sounds interesting, by all means write to the company directly. Only the company itself can provide final, authoritative information about its products and prices. Don't hesitate to write to more than one company, if more than one company provides the specific product in which you are interested. In this way you can compare before you buy.

COMPANY INDEX

SOLAR-PAK
P.O. Box 23585
Tampa, Fla. 33622

PRODUCTS:

solar water heater plans

Plans prepared by this company are designed for home use as an auxiliary unit to a new or existing hot water system. The system can be modified for other uses, such as snow and ice removal for walks and driveways, or as a prime source for places with large storage capacities—car washes, restaurants, school cafeterias, etc.

Plans, material and tool list $7.00

COMPANY INDEX

AIRBORNE SALES CO., INC.
8501 Steller Dr.
Culver City, Calif. 90230

D.C. Generator

PRODUCTS:

AC-DC generators
actuators, 12V
actuators, 24V
air compressors, 12V
arc welders
electric valves, 24V
fans, 12V
fans, 24V
gear trains & motors
generators
lights, 24V
motors, 12V
motors, 24V
power suppliers
pressure switches
regulators, 12V
relays, 12V
reversing relays, 24V
water pumps, 6V
water pumps, 12V
water pumps, 24V
winches, 12V

Aircraft electrical and electronics components, machine tools, hardware, survival equipment (aircraft and marine) and

AIRBORNE SALES CO., INC. (Cont'd)
government surplus is supplied by this company. It won't matter if you want a leach relay or a log-pulling winch that operates off of 12 volts or a generator that puts out both 24 volts DC at 30 amps and 120 volts AC at 9 amps at the same time, this is one of the best places to start looking. The company has been in business since 1945 and has some of the best bargains on the market. They even have low-voltage actuated valves that will turn the water running into your water wheel on or off so you'll never have to leave the comfort of your home in the winter to start your electricity going. If you don't have a water-powered generator, there are over 50 different ones here, with voltage ranging from 6 to 220 volts, that could easily be attached to a water wheel, or even to a propeller to utilize the wind.

The 88 page catalog of wonderful things is only 50¢

AIRCRAFT COMPONENTS, INC.
P.O. Box 1188
Benton Harbor, Mich. 49002

2½ Ton Winch, 12 Volts

PRODUCTS:

air compressors, 12V
batteries, 12V
batteries, 24V
fluid pumps, 12V
incandescent lights, 12V
winches, 12V
wind direction indicators
wind speed indicators
wind speed indicators, portable

This is a mail order house for aircraft parts and supplies. Batteries are available in both rubber and metal cases and from 12 to 24 volts with ratings up to 90 amp hours. Batteries for aircraft usually have a heavier construction, longer run-down time and greater surge capacity than most automotive batteries. They also have wind speed and direction indicators ranging from small hand-held units to very refined instruments costing several hundred dollars. The winches

AIRCRAFT COMPONENTS, INC. (Cont'd)
they carry are designed to pull aircraft out of hangers and have capacities from 2800 to 4600 pounds up 20% grades.

Catalog that is prepared like a good magazine free

AMERICAN WIND TURBINE
P. O. Box 446
St. Cloud, Fla. 32769

PRODUCTS:

wind generator plans

This company has one of the more promising designs for blade design now available. The hub is made of a bicycle wheel and the complete unit is light enough to be handled by one person with ease. The unit has been claimed to operate at a 50% efficiency.

Information on request

AUTOMOTIVE STEAM SYSTEMS
8591 Pyle Way
Midway City, Calif. 92655

PRODUCTS: *Steam-Powered Sports Coupe*

air atomizing burners
autos, steam powered

AUTOMOTIVE STEAM SYSTEMS (Cont'd)

bicycles, steam powered
motorcycles, steam powered
plans, steam buggy
steam engines
steam generators
steam generator plans
steam throttles

The latest designs, technology and information on the state of the art in steam development will be found in the Automotive Steam Systems catalog. You'll find information on steam engines, steam generators, valve components and auxiliary equipment listed for those who wish to buy ready made, custom equipment. For those who would rather be do-it-yourselfers they offer plans, materials and castings that will allow you to save quite a bit of money. To do it yourself you should have available to you a lathe, milling machine, drill press and other machine shop tools. Steam throttles are made in three bore sizes out of brass or stainless steel and operate with a quarter turn in either direction to fully open. The steam engines are Mercury outboard 4 and 6 cylinder blocks with the reed valves removed and cast plugs made and installed in the exhaust ports. Additional custom work available is to remove water jackets and install stainless steel covers or make and install large size impulse valve plenum chambers. The steam generator plans are put into a comprehensive set of information comprised of four large (C size) blue line prints plus four smaller sheets enabling any skilled mechanic to convert certain internal combustion engines to steam, with details on engine valving, exhausting, and a simple lubricating system. The air atomizing burner nozzle was developed specifically for the purpose of aiding anyone in obtaining a clean-burning, smog-free fire in the combustion chamber. A bicycle and motorcycle powered by small steam engines that use propane for fuel are available in addition to a small sports car using a 4 cylinder steam engine that burns kerosene or diesel fuel. Steam buggy plans are available for building an original styled horseless carriage replica of the

1900's that only weighs 500 pounds and is capable of speeds up to 35 mph. These plans include boiler plans and directions for aligning the tiller steering and trailing arm suspension.

Illustrated and fun to read catalog 50¢

ALTERNATIVE ENERGY SYSTEMS
LES AUERBACH
242 Copse Rd.
Madison, Conn. 06433

PRODUCTS:

methane publications

"A Homesite Power Unit: Methane Generator" is a 50 page book by Les Auerbach that explains a methane generator's operation and construction in terms that a layman can understand and follow. He takes the reader from the planning stages of various temperatures for operation through mixing ratios for proper gas production, through gas extraction, through design of digesters and storage for the resulting gas. Design examples are given for units that are presently producing, such as the unit at the University of California at Berkeley, which was designed as a model system that would demonstrate the feasibility of methane and compost production and recycling of organic wastes. The system is composed of a digester with loading apparatus, a gas storage facility, a pressure gauge apparatus, a gas meter and a gas cooking stove. The description of the model includes schematic cutaways, construction details and operating procedures that could be used to build an identical system.

A well illustrated book with a wealth of information that is well worth the $5.00 price

AUTOMATIC POWER DIVISION
P.O. Box 18738
Houston, Tex. 77023

PRODUCTS:

generators, wind powered

Automatic Power industrial "Aerowatt" wind generators are available in five sizes from 28 to 4100 watts. The machines are sturdily constructed to prevent damage in high winds and to insure trouble-free unattended operation. The generator housings are machine-cast aluminum, and all hardware and ferrous parts are suitably plated to resist corrosion. The generator is a brushless three phase AC permanent magnet type which eliminates the usual problems associated with brushes. The rotational speed of the propeller is regulated by a unique centrifugally-operated pitch control system. In higher wind velocity, the variable pitch control maintains a constant speed by changing the angle of attack of the blades while limiting the rotational speed to a safe value in winds up to 125 mph. The blades are machined hardwood and are carefully balanced to eliminate vibration. Hardwood is used because the ultimate strength-to-density ratio is better than most titaniums, aluminum or ordinary steels. Galvanized towers can be furnished which are custom designed to meet the requirements of the site and may be self-supporting or guyed with steel cable. To further assure the optimum use of low wind velocities, the governing system is designed to move the blades to a starting position when rotation stops. This insures maximum starting torque in low wind velocities. These features make the Aerowatt generators particularly useful in areas of low annual wind velocity.

Catalog illustrated with plans and schematics is free

THEODORE BARGMAN CO.

129 Industrial Ave.
Coldwater, Mich. 49036

PRODUCTS:

fluorescent lights, 12V   incandescent lights, 12V

All of these products are manufactured for recreational vehicles, and they are all low voltage, low amperage, and would work off of any small generating system made with automobile parts. All the lights use between 10.8 and 14.5 volts DC and have a reverse polarity safeguard built in.

Catalogs and brochures are free

BRACE RESEARCH INSTITUTE

MacDonald College of McGill University
Ste. Anne de Bellevue 800
Quebec, Canada

PRODUCTS:

solar publications
wind publications

Solar Water Heater

The Institute has developed several low cost machines that are designed for underdeveloped countries with a low technology level, and have produced booklets giving plans and directions for building these units. A simple solar steam cooker consists of a central pipe which is connected to a reservoir and is placed in an insulated box covered by two or three thin glass sheets. The top of the pipe is connected to an insulated basin containing the cooking pot, where the steam condenses from the basin and produces enough heat to prepare boiled and stewed foods. Their Savonius rotor windmill pump is made of two 45 gallon drums cut in half lengthwise and mounted in a vertical frame with the shaft connected to a diaphragm pump via an eccentric and rod assembly. This windmill can pump up to 300 gallons of water per hour under 10 feet of water pressure, or be used to power a generator, or both. They also have booklets on how to make a solar still, a solar swimming pool heater and a solar water heater from wood and plastic, another type of solar water heater made from glass and concrete, and a solar cabinet dryer for agriculture produce. All the booklets are written in simple language with easy-to-follow diagrams, and are in four different languages.

List of publications is free

BUDGEN & ASSOCIATES
72 Broadview Ave.
Pointe Claire, Quebec
Canada

Wind-Powered Generator

PRODUCTS:

generators, wind powered
solar publications
wind publications
wind pumps

For Canadians, here are the agents for the Lubing wind generators and wind pumps. Also the Elektro GMBH wind generator from Switzerland. They also manufacture a 32 foot diameter windmill designed by the Brace Research Institute for both pumping and electric generation. The Lubing machine will generate up to 400 watts in wind speeds of 26

BUDGEN & ASSOCIATES (Cont'd)

mph, and the Elektro machine is available in sizes from 50 to 12 kw. Dr. H. P. Budgen is also the chief consultant for the Brace Research Institute and has been instrumental in many of their research projects and developments. The company also handles publications for the Institute.

Packets and brochures which help you to choose a system that is proper for your needs are available at nomical costs which will be quoted on request

CALDWELL INDUSTRIES
Box 170
Luling, Tex. 78648

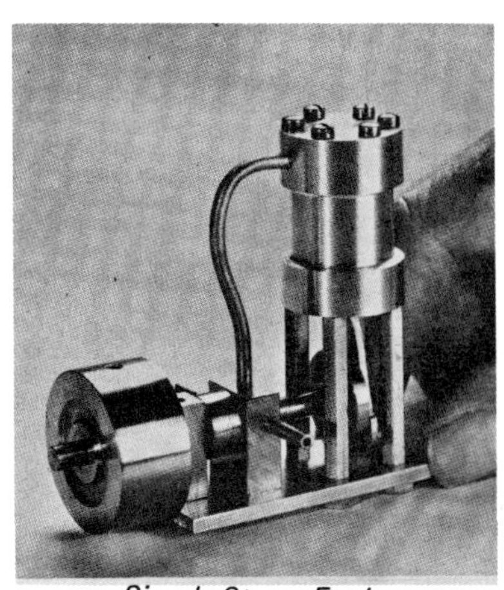
Simple Steam Engine

PRODUCTS:

- steam engines
- steam engine kits

Most of the steam engines and kits from this company are models, but some can be utilized for power needs. Most of the engines, however, are sold as castings and require machining to be assembled. This is done so that the prices are within reason. Some engines are available ready to be assembled. You can obtain single and multiple cylinder engines that produce up to 2/3 h.p., which would be sufficient to drive a small generator of about 250 watts or run a pump.

The catalog is a bit hard to read but is full of pictures and descriptions to help you make your choice and is available for $1.00

CRYTON OPTICS, INC.
7 Skillman St.
Roslyn, N.Y. 11576

PRODUCTS:

fresnel lenses

Cryton is a manufacturer of fresnel lenses, both cylindrical and with concentric grooves, for use in solar energy.

Catalog free

CULLMAN WHEEL CO.
205 Huehl Road
Northbrook, Ill. 60062

PRODUCTS:

chains and sprockets

Cullman Wheel Company is a manufacturer of roller chain sprockets and a supplier of roller chain and roller chain parts. Their products are serviced through the Northbrook, Ill., factory warehouse and eight warehouse locations throughout the country. Their roller chain sprocket line is unique in that in most of the small and middle size pitches of chain they offer sprockets of consecutive numbers of teeth in "A" style plate sprockets, "B" style plain bore sprockets, as well as finished bore sprockets complete with keyways and set screws. This company can supply just about any size, from a 9 tooth 1-1/8 inch diameter sprocket with a 3/8 inch shaft hole to an 80 tooth 52 inch diameter sprocket with a 3 inch shaft hole.

Catalog free, 111 pages of illustrations and specifications followed by 32 pages of tables and engineering data

DAVIS INSTRUMENTS CORP.
857 Thornton St.
San Leandro, Calif. 94557

PRODUCTS:

solar cookers

Solar Cooker

The solar cooker uses only the sun's heat to cook the meal from a pollution-free fuel. It is ideal for campers, picnickers or homesteaders who have a fuel problem. You just wrap skewered food in special Solar Foil included with the cooker, aim it with the tracking cell, and in 10 to 12 minutes the food will be done.

Catalog free

DEMPSTER INDUSTRIES, INC.
P.O. Box 849
Beatrice, Neb. 68310

PRODUCTS:

towers wind pumps

Dempster manufactures a complete line of water systems. On their new Annu-Oiled No. 12 windmill you get a five year warranty against defects and workmanship. They have machine-cut equalizing gears that work independently of each other on a steel bearing. The wheel can come loose because the wheel spider or retainer is made with a split hub which clamps around the wheel shaft. The machine only

DEMPSTER INDUSTRIES (Cont'd)

requires oiling once a year, at which time you drain out the old oil, flush with kerosene, and refill with new oil. All the towers have wood platforms to insure safety and accessibility. They are available from 22 feet to 39 feet and feature girts every 5½ feet.

A super fun to read and browse catalog is free for the asking

DESERT SUNSHINE EXPOSURE TESTS, INC.

Box 185 Black Canyon Stage
Phoenix, Ariz. 85020

Solar Insulation Instrumentation

PRODUCTS:

solar laboratory

This is an outdoor weathering laboratory specializing in accelerated intensified exterior exposure of plastics, paints, glass and textiles. It is the only facility equipped to accelerate natural weathering utilizing solar energy itself. They also provide independent testing and evaluation of solar energy developments for qualified companies and individuals.

Write for catalog, stating requirements

DUBIN-MINDELL-BLOOME ASSOCIATES P.C.
42 West 39th St.
New York, N.Y. 10018

PRODUCTS:

solar consultation   solar design   solar engineering

This company is geared more towards commercial and industrial applications for government agencies and companies involved in research and development. They have done research and feasibility studies for the application of solar energy, wind energy and methane power for a residential housing project, Grassy Brook Village, in Brookline, Vermont, a condominium that derives its energy from natural and non-polluting sources. The firm was founded in 1946 with a philosophy to first consider the environment for those who use the facility, analyze the types of environmental control systems which are most appropriate, research materials, determine energy requirements and initial operating and maintenance costs. They also have an overseas office in Rome that is responsible for the company's projects in Europe and Africa.

No catalog but they have several pamphlets that show larger installations that could be scaled down for residential use

J. A. DUFFIE
University of Wisconsin
Engineering Experiment Station
1500 Johnson Drive
Madison, Wisc. 53706

PRODUCTS:

solar energy courses

John A. Duffie is the director of the solar energy laboratory at the University of Wisconsin and offers courses in solar energy. The courses cover solar radiation, solar flat-plate

J. A. DUFFIE (Cont'd)

collectors, flat-plate collector design and performance, solar focusing collectors, energy storage, solar heating and cooling, and many other topics. The major objective of the courses is to provide engineers with a working knowledge of thermal processes for solar energy utilization. The courses are held on a regular basis every six or seven months and last for five days. The tuition fee is $350.00.

More information can be obtained by writing to the school

DWYER INSTRUMENTS, INC.
P.O. Box 373
Michigan City, Ind. 46360

PRODUCTS:

wind speed indicators

Two types of wind measuring instruments are manufactured by Dwyer. The smallest is a hand-held wind meter that has a dual range of 2 to 10 mph or 4 to 46 mph. It is rugged enough for sea-going use but still retains pocket-size convenience. The second model consists of a wind speed indicator that can be mounted on an interior wall, a rooftop-mounted vane and sensor, and 50 feet of flexible double

DWYER INSTRUMENTS, INC. (Cont'd)

column tubing. The unit is 100% pneumatic or "wind powered," and operates on the same principle as aircraft speed indicators.

Illustrated product sheet free

E. I. & I. ASSOCIATES
P.O. Box 37
Newbury Park, Calif. 91320

PRODUCTS:

solar publications

E.I. & I. Associates publish an illustrated guide that gives the basic component and system design considerations required of flat-plate solar collectors for liquid/air heat transferring applications. The book explains basic costs, energy basics, collector construction, flat-plate design, plumbing hook-ups, storage designs, and gives tables for latitudes across the U.S. Solar water heating for faucet hot water applications has been principally stressed, because it is within the realm of the average home owner's ability and budget. Solar water heating for space heating applications, however, requires a collector size and storage capacity of such magnitude that they feel it is unrealistic to consider it seriously. Over half of the 24 page booklet is devoted to various graphics concerned with design and application, showing cross-sections of collectors, collector connection arrangements, fluid tube patterns, circulation systems, etc. Anyone seriously considering building his own solar water heater would do well to consult this book.

Illustrated Solar Energy Guide: of Flat-Plate Collectors for Practical Home Applications

EARTHMIND
26510 Josel Dr.
Saugus, Calif. 91350

PRODUCTS:

methane plans
methane publications
solar publications
solar water heater plans
wind generator plans
wind publications

Earthmind is a small group of people living on a farm and operating a non-profit (state and federal) research and educational program, publishing their findings at a price that just about anyone can afford. One of the books that they have published is "Wind & Windspinners," a nuts and bolts approach to wind electric systems that delves mainly into the S-rotor type wind generator. The book starts out with the fundamentals of wind energy and electricity generation, and covers batteries (their selection and operation), different types of control mechanisms, and the final design and principles of the S-rotor wind generator. This group also has equally good publications on solar energy and its utilization, as well as methane and its uses. These books and plans are the result of actually working with and developing equipment so that others might duplicate similar equipment at low cost.

The publication list is sent free, but a self-addressed, stamped envelope should be sent along with the request

EDMUND SCIENTIFIC CO.
555 Edscorp Bldg.
Barrington, N.J. 08007

Parabolic Mirror

PRODUCTS:

air compressors, 12V
generators, wind powered
Mylar
solar cigarette lighters
solar cookers
solar furnaces

parabolic mirrors
solar cell banks
solar cells
solar greenhouse plans
solar house heater plans
solar water heaters
wind speed indicators

This company offers over 4500 different products, many of them useful for alternative energy use (from 12 volt air compressors to wind speed indicators). You can get a solar cooker to cook your food, a small parabolic mirror to light your cigarettes, wind generators, a 2000 degree solar furnace, and many other items at low prices. Solar cells are sold both in kit form and in completed banks so that you can adapt them to any use requiring a low voltage input. Every product in their catalog seems to be described as honestly and accurately as possible. In addition, if you are not completely satisfied with any purchase, return it within their 30 day limit and the company will refund your money.

A real fun to browse catalog is free to those who ask

ELEKTRO GMBH
St. Gallerstr. 27
8400 Winterhur
Switzerland

PRODUCTS:

generators, wind powered

5 KW Generator

If it is a question of solving a special problem, such as the supply of energy to a mountain station or an isolated farm, a radio station or a meteorological or biological station, the installation of a wind-driven generator will be reasonable. Depending on its situation, the fitting of the wind-driven generator has to be considered on a 20 to 60 foot tower.

ELECTRO GMBH (Cont'd)

This Swiss company can supply wind plants from 150 up to 12,000 watts, all of which have brushless permanent magnet rotors. An optional automatic control replaces the hand brake on some models and is mounted at the foot of the tower. This device turns off the generator on a fully loaded battery and in wind speeds of 50 to 70 mph, depending on the model. Some of the units from this company have been in constant use for 20 years in the high mountain regions of Europe without any severe maintenance problems.

Illustrated brochures free

ENERGY TRANSFER SYSTEMS

204 W. 13th St.
Lawrence, Kan. 66044

PRODUCTS:

methane publications   solar publications   wind publications

This company offers assistance and information concerning wind-generated electrical systems, solar house heating and methane digestors. They also offer site analysis for wind generator and solar projects underway.

No catalog

ENVIRONMENTAL ENERGIES, INC.

21243 Grand River Ave.
Detroit, Mich. 48219

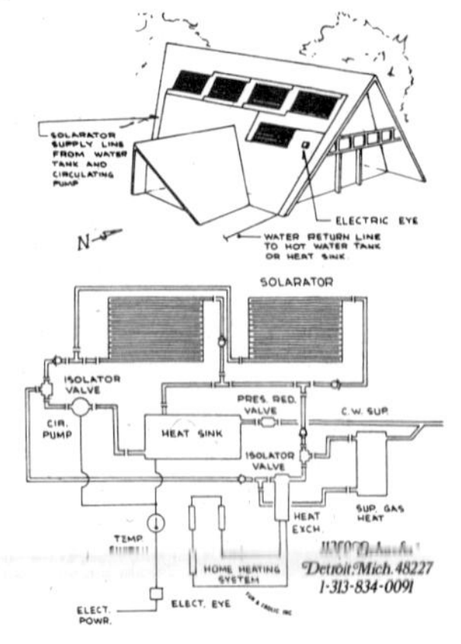

PRODUCTS:

- batteries
- generators, wind powered
- inverters
- solar cell banks
- solar collectors

ENVIRONMENTAL ENERGIES, INC. (Cont'd)

This is one of the newer companies set up to research, integrate and market alternative non-polluting energy systems that are compatible with residential living. They sell wind generating systems from 12 volt 200 amp units up to 115 volt 4500 watt commercial units, complete with controls and accessories for special adaptations on all models. The batteries have individual flow charge indicators and are encased in clear polystyrene. Specially selected separators resist heat and allow full circulation of the electrolyte, while a specially compounded insulation prevents shedding of active material. The inverters are available in 1000 to 4000 watt capacities with variable cycle adjustments and operate with a 95% efficiency. Both the solar collectors and the solar cell banks marketed by this company appear to be of high quality workmanship and designed for extended usage.

For $2.00 you receive their catalog, which is full of useful information in addition to their own product line

FAFCO, INC.
138 Jefferson Drive
Menlo Park, Calif. 94025

PRODUCTS:

solar water heaters

FAFCO, INC. (Cont'd)

Fafco manufactures solar heat exchangers for heating swimming pools and other low temperature water heating applications. Complete automatic control systems are available for the solar heaters. Fafco has been manufacturing, selling and installing solar heaters for over three years. They claim their solar heat exchangers are the first economically viable and cost competitive solar heating systems to be manufactured on a mass production basis. Also they claim that their solar pool heater pays for itself in two to five years with the money it saves on the fuel bill.

A brochure and question and answer booklet is sent free

FESCO (Fred E. Stewart Co.)
(formerly Bucknell Engineering Co., Inc.)
P.O. Box 3426
El Monte, Calif. 91733

PRODUCTS:

alternators
generators
generators, motor powered
generators, wind powered
wind generator parts

This company does all of its own mechanical and electrical engineering; makes its own engineering drawings, parts lists, manuals and specification sheets; does most of its own tooling, all of its own machine work, winding, mechanical and electrical testing as well as manufacturing all of its own control components such as transformers and solid state circuits. The motor powered generators can be fueled with either gas or diesel and have output ranges from 1 kw to 25 kw, 60 or 400 cycle, and single or three phase windings (or multiple windings for combinations or output voltages). A wind generator is also manufactured that was originally designed to furnish a small amount of electrical power for remote, unat-

FESCO, INC. (Cont'd)

tended stations and be rugged enough to withstand storms. It is a 14 pole, permanent-magnet type rotating field generator directly connected to a propeller. The total weight of the generator assembly is about 140 pounds, and it begins to charge a 12 volt battery with a 7 mph wind. At 9 mph it charges at 1 amp, at 13 mph it charges at 3 amps, increasing to 5 amps at 15 mph and 10 amps at 20 mph. The maximum charge rate is 20 amps, and if the wind increases beyond 35 mph the complete generator assembly tilts back and lets the wind begin to spill off the propeller. The tilting is controlled by a hydraulic cushioning device that will operate from -65 deg. F to 155 deg. F.

$3.50 is charged for the wind generator information packet, which includes a cover letter, technical write-up, wind velocity chart, drawing of unit, and a company brochure.

FILON

12333 S. Van Ness Ave.
Hawthorne, Calif. 9025(

MATERIAL	PERCENT TRANSMITTANCE OF:		
	Light Energy	Total Solar Energy	Thermal Energy
Glass	90	73.4	4.4
Filon Panel	95	84.8	1.0

PRODUCTS:

solar collector covers

Filon, a division of Vistron Corporation, manufactures translucent fiberglass-plastic panels used in solar heat collectors and in greenhouses to conserve fossil fuel. They are called Filon Greenhouse Panels. A small-scale test of a simple solar heat collector has shown that Filon panels are more efficient than glass covers in raising water temperature. The test compared 3/16" glass sheets with Filon panels that were one-sixth as thick. The top of the line is the "supreme" grade that gives you a 20 year full replacement guarantee with 92% light transmission. The next grade below that is the "superior"

FILON (Cont'd)

grade with a 15 year full guarantee for replacement and 92% light transmission properties. They offer the "standard" grade with a 10 year guarantee that is pro-rated over that period, and the "economy" grade that has a pro-rated 5 year guarantee and 85% light transmission. You can buy any of these four grades in either corrugated or flat sheets or in 100 foot rolls.

Catalog free

FORRESTAL CAMPUS LIBRARY
Princeton University
Princeton, N.J. 08540

PRODUCTS:

wind publications

In a search for electric power from the wind, Princeton has developed a sailwing wind turbine.

Information is available upon request

JERRY FRIEDBERG
ARRAKIS VOLKSWAGEN
Box 531
Point Arena, Calif. 95468

Propane Powered Truck

JERRY FRIEDBURG (Cont'd)

PRODUCTS:

- methane publications
- propane, auto conversion kits
- propane, auto conversion parts

Here, apparently, is the only company in the country that provides propane and methane conversion equipment at a discount. They will provide anything wanted in the way of propane or methane conversion equipment at 1/3 of list price. They are also the only place for do-it-yourselfers, enabling people to convert their autos for $100-$200 instead of the $500-$800 charged by conventional conversion places. The company claims these conversions will work on any internal combustion engine from generators to Volkswagens to Cadillacs and tractors. Some of the components offered are convertors, fittings, high pressure hose and connectors, vapor hoses, gasoline on-off valves, manual primer wires and other hardware. Methane works with all the kits sold by this company, but they add, "It works, but is more hassle, needs more equipment to produce the gas, and lots of manure." The manual sold by this company tells how it is done and who else to buy from if you don't like them or their kits.

Illustrated brochure is free, and the complete manual is $2.00

L. JOHN FRY

1223 North Nopal St.
Santa Barbara, Calif. 93103

PRODUCTS:

methane publications

Here is the first practical book on how to design and build your own displacement-type methane generating plant. Authored by the pioneer and innovator of the first, continuous-

L. JOHN FRY (Cont'd)

ly operated displacement digestor methane plant, the book is written for operators of small and large farms, homesteads, dairies, canneries, etc., and is intended to foster rural energy independence. Included are numerous charts, diagrams and photos showing how to produce and use your own supply of free energy and fertilizing material retaining nature's full nutrient value. Important chapters deal with the biology of digestion, raw materials, gas and gas uses, and various sludge uses. The book is unique in that nothing has ever been published before on the subject of continuously operated displacement digesters of horizontal design, particularly offering solutions to the scum removal problem, a major source of difficulty with methane power plants. The three main types of methane power plants (batch load, vertical and horizontal) are covered, as well as everything from inexpensive working models to farm-integrated power plants. The book deals clearly and succinctly with technical aspects of the biological process and the raw materials used.

"Practical Building of Methane Power Plants" by L. John Fry is $12.00

GAYDARDT INDUSTRIES
Rt. 1 Box 319-A
Brandywine, Md. 20613

PRODUCTS:

solar collectors

Gaydardt manufactures a collector that is 4′ x 8′ x 6″ in a 26 gallon box that does not have to be at a 45 degree slope because of its unusual design. It can be modified for home heating in conjunction with existing heating plants and it may be used with radiant heat, baseboard heating or forced air heating. Although supplemental, the reduction of fuel consumption would soon compensate for the cost on installation. They claim their collectors will give a 5 degree pickup at 35 gallons per minute.

Catalogs are free of charge

GOBAR GAS RESEARCH STATION

Ajitmal
Etawah
(U. P.)
India

PRODUCTS:

methane plans   methane publications

This is Ram Bux Singh's project headquarters where he is publishing some very informative books. The first book, "Bio-Gas Plant, Generating Methane from Organic Wastes," deals with methane principles, design, and useage of the final product, in addition to the history of bio-gas production. He has researched, designed and built equipment for use in both hot and cold climates and offers the results in these books. The second book, "Bio-Gas Plant Designs with Specifications," gives various types and sizes of methane systems ranging from 50 cubic feet of production (family size) to 10,000 cubic feet of gas production (industrial size). They have been successfully demonstrated and put into use throughout the world since 1960. Valuable data about digestion cycle, temperature control, bacterial activity, analysis of compost and agitation systems have been compiled into these two bio-gas research manuals and make them a necessity for anyone thinging about or building a methane system. For the price of these books

GOBAR GAS RESEARCH STATION (Cont'd)

(which barely covers the cost of mailing from India), there is no other set of books that offers as much information to the individual.

"Bio-Gas Plant," 105 pages of plans and data, is $5.00
"Bio-Gas Plant Designs with Specifications," 49 pages of plans, is $7.00

GRAY COMPANY ENTERPRISES CORP.
7701 N. Stemmons Freeway No. 245
Dallas, Texas 75247

PRODUCTS:

steam generators

The Gray vapor generator represents the first major advance in steam technology in over 200 years. This small generator (14 inches high and 7 inches in diameter, with a weight of 65 pounds) is capable of producing one million BTU's of heat per hour when utilizing a mixture of gas and air (hydrocarbons). When fueled by hydrogen and oxygen, the Gray Vapor Generator is capable of producing three million BTU's per hour without any pollutants. Because of the lack of moving parts, little servicing is required, and it is able to use any fuel more economically than its competitors, it is claimed, because of the excellent thermal design characteristics. Anywhere steam is required, this little generator will do the job. The company says it will work with both methane and natural gas with great efficiency as well.

Catalog free upon request

HOT WATER
647 Chelam Way
Santa Barbara, Calif. 93108

PRODUCTS:

solar publications

"Hot Water" is a small booklet by Scott & Chole Morgan and David & Susan Taylor written and illustrated to provide the reader with simple, inexpensive home construction plans for building solar and stack coil water heaters. Simplified drawings also show how to adapt these systems to water heating systems already installed.

Booklet with plans and illustrations $2.00

INTERMEDIATE TECHNOLOGY DEVELOPMENT GROUP
Parnell House
25 Wilton Road
London, SW IV 1JS
England

PRODUCTS:

methane publications

The Intermediate Technology Development Group, London, was established in 1965 to investigate ways and means of utilizing to the fullest extent the resources available to developing countries through the application of appropriate technologies. "Methane: Fuel for the Future" is one of the books put out by this group that surveys methane gas around the world, giving diagrams, photos and copious references.

Catalog of publications is free

KALWALL CORP.
1111 Candia Road
Manchester, N.H. 03105

PRODUCTS:

solar collector covers

This is a durable, efficient fiberglass cover for a heat collector that is more versitile than glass. Transmittance properties are similar to glass, but it is unbreakable.

Catalog free

KOHLER CO.
Kohler, Wisc. 53044

PRODUCTS:

engines, gas powered

generators, gas powered

3 KW, 120 Volt Generator

Kohler manufactures 4-cycle cast iron engines ranging from 4 to 24 horsepower and also engine-driven generator sets ranging from 500 watts to 500 kilowatts. The engines, from the smallest to the largest, are solidly built for heavy duty use and a long, trouble-free life. They are designed with a large bore and a short stroke to give higher rpm's. They contain oil-bathed governors, automatic compression releases, and dry element air cleaners. Kohler has been making generator sets since 1920 and has had units on South Pole expeditions, desert wastelands and many other diverse places to prove their stability. Some of the smaller units can be operated on as little as ¼ of a gallon of gas per hour under full load and produce 120 volts and 750 watts.

Brochures are free

JAMES LEFFEL & CO.
Springfield, Ohio 45501

PRODUCTS:

- hydro-electric plants
- water wheels

Leffel builds a complete line of turbines, from a fraction of a horsepower up to a very large capacity. They also sell water wheels. The Hoppes Hydroelectric Unit, a smaller type, built in capacities of up to 10 kilowatts and for heads up to and including 25 feet, is shown in their Bulletin H-49. This bulletin tells not only the kilowatt capacity of each unit but also the kilowatt output in each instance, as well as the quantity of water and the speed. If you write to them you should also request a copy of Pamphlet A, which describes how to measure stream flow in terms of cubic feet per minute and how to measure head. With head and quantity of water, you can compute the power which can be developed, in terms of kilowatts. The smaller units, such as the HL model, can produce half a kilowatt with only an eight foot head of water, or you can buy a 10 kilowatt unit that requires a 25 foot head to operate properly. This company has been in operation for over 112 years, and the workmanship of its units is proven by the expected operating lifetime without repairs, which is in the 90 to 100 year range.

Brochures are free

THE MOTHER EARTH NEWS
P.O. Box 70
Hendersonville, N.C. 28739

PRODUCTS:

publications

"The Mother Earth News" contains articles in almost every issue concerning alternative energy sources and how to build the hardware to use the sources, with photographs and copy written by the experimenters themselves. In addition, they have a research and development department working primarily in the field of energy. They have also done experimental work in methane production and have published a paperback book called "Mother's Handbook of Homemade Power," which is $1.95 and describes possible ways of using wood, solar, methane and wind energy at home.

Subscriptions are $8.00/yr for 6 issues

MOTOROLA AUTOMOTIVE
& INDUSTRIAL PRODUCTS
9401 W. Grand Ave.
Franklin Park, Ill. 60131

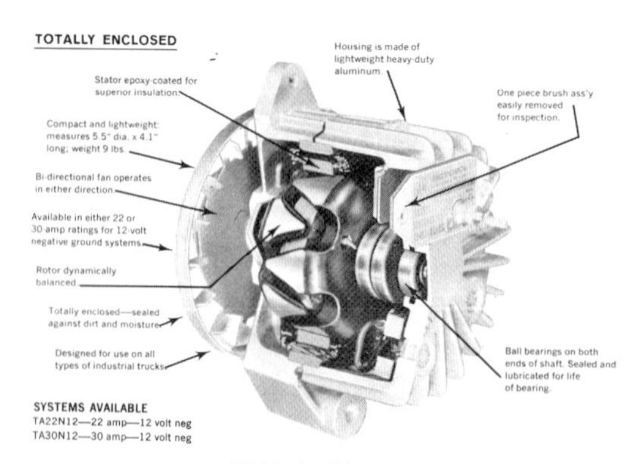

12 Volt Alternator

PRODUCTS:

alternators

MOTOROLA AUTOMOTIVE (Cont'd)

This company offers high quality alternators at relatively low cost. Their alternators range from 6 through 32 volts, with a 6 volt/55 amp unit as the smallest of the line, up through a 12 volt/120 amp unit, a 24 volt/70 amp unit, and up to the 32 volt/70 amp unit designed for heavy industrial use. These alternators are worth looking into if you are planning to build your own electrical generating system from components, because several of them have special provisions for a low speed cut-in that would take advantage of a slow wind or dropping water flow.

Catalogs and information sheets are free

NATIONAL AERONAUTICS & SPACE ADMINISTRATION (NASA)

Technology Utilization Office
George C. Marshall Space Flight Center
Marshall Space Flight Center, Ala. 35812

PRODUCTS:

solar publications

Even though the information released through NASA's reports is of a highly technical nature, many good ideas and information can be derived from them. One such report is called, "The Development of a Solar-Powered Residential Heating and Cooling System," and describes the efforts to demonstrate the engineering feasibility of utilizing solar power for residential heating and cooling. These efforts have been concentrated on the analysis, design and test of a full-scale demonstration system which is currently under construction in Huntsville, Alabama. The basic solar heating and cooling system utilizes a flat plate collector, a large water tank for thermal energy storage, heat exchangers for space heating and water heating, and an absorption cycle air conditioner for space cooling.

Catalog of publications is free

NATIONAL BUILDING RESEARCH INSTITUTE
P.O. Box 395
Pretoria
South Africa

PRODUCTS:

solar publications

This is a research organization, and as such does not market any particular product. They do, however, have several publications in the solar energy field that are well worth obtaining. Research Report No. 248 is an 80 page book on construction and useage of various types of solar collectors for use in Africa. They give several tables giving comparisons of the different collector designs and also diagrams and photographs showing the collector units in various stages of construction and use.

Report No. 248 costs $3.00

THE NATIONAL CLIMATIC CENTER
Asheville, N.C. 28801

PRODUCTS:

solar publications wind publications

Here you can get wind and solar data for the entire U.S. The maps they sell show the average amount of sunlight per day and the average amount of wind per day, and also the usual direction of wind travel.

NEGEYE LABORATORIES
P.O. Box 547
Pennsboro, W. Va. 26415

PRODUCTS:

generators, wind powered
hydrogen generator plans
wind generator manuals

Negeye already has engines running on hydrogen and is selling plans for the generator and conversions to enable you to run your car, tractor or truck on hydrogen. They also have out-of-print books that they are reproducing, such as the Fairbanks-Morse Home Lighting Plants (1914), or the Delco Manual for their 850/1250 watt kerosene plants, the Wincharger Manual for the 650 watt, 32 volt units, and a booklet on rebuilding 32 volt batteries. They also sell Wincharger wind generator systems and accessories.

The catalog costs $1.00, which also puts you on the mailing list to receive new information as it is released

THE NEW ALCHEMY INSTITUTE
Box 432
Woods Hole, Mass. 02543

PRODUCTS:

methane publications
solar publications
wind publications

The Institute is a small, international organization for research and education on behalf of man. It is a non-profit, tax-exempt group, and derives its support from private contributions and research grants. Among its major tasks is the creation of ecologically-derived forms of energy, agriculture, aquaculture, housing and landscapes that will permit a revitalization and repopulation of the countryside. They have five publications on windmills, among them, "A Windmill Bibliography," prepared by Marcus Sherman, which gives sources for home built and commercial water pumping and electricity generating windmills. It also gives titles of books about windmills and lists organizations involved in their development and manufacture. This is free, but they ask that you send a self-addressed, stamped envelope with your request. Another windmill book is "Electronics for Homebuilt Windmills" by Fred Archibald. This gives you a comprehensive and valuable description of do-it-yourself electronics for windmills. Two methane publications are offered, one of which is titled "Methane Digestors for Fuel, Gas and Fertilizer," by L. John Fry (q.v.) and Richard Merril. This gives a very valuable treatment of methane systems and research, including designs for a small and an intermediate scale system, and is available for $3.00 from NAI-West, 15 West Anapamu, Santa Barbara, Calif. 93101. The solar publication that they offer is "The Ark," which is about design and rationale of a solar heated greenhouse and aquaculture complex adapted to northern climates.

An Associate Membership in the Institute is $25.00 per annum

NOVA ELECTRIC MFG. CO.
263 Hillside Ave.
Nutley, N.J. 07110

PRODUCTS:

- frequency changers
- inverters

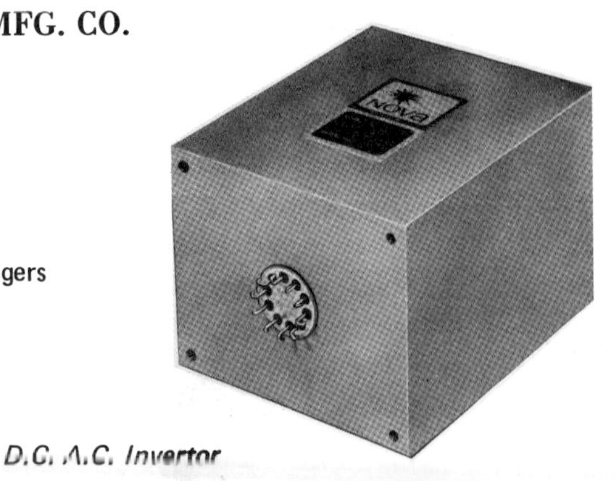

D.C. A.C. Invertor

NOVA ELECTRIC MFG. CO. (Cont'd)

Nova offers over 1,000 three-phase and single-phase inverters and frequency changers at low catalog prices. Input ranges are from 11 to 350 volts DC, with output ratings from 125 to 3,000 volts AC single phase and 150 to 9,000 volts AC three phase, and output frequencies from 48 to 440 Hz continuously variable or 50, 60 and 400 Hz fixed. These units are largely maintenance-free, and of rugged industrial construction for continuous operation at a 100% duty cycle. All units also contain a short circuit and a reverse polarity protection system and are built to withstand temperatures from -20 degrees C to +50 degrees C with no drop in efficiency.

Catalog is free

ONAN DIVISION–
ONAN CORP.

1400 73rd Ave. N.E.
Minneapolis, Minn. 55432

PRODUCTS:

- generators, gas powered
- generators, PTO powered

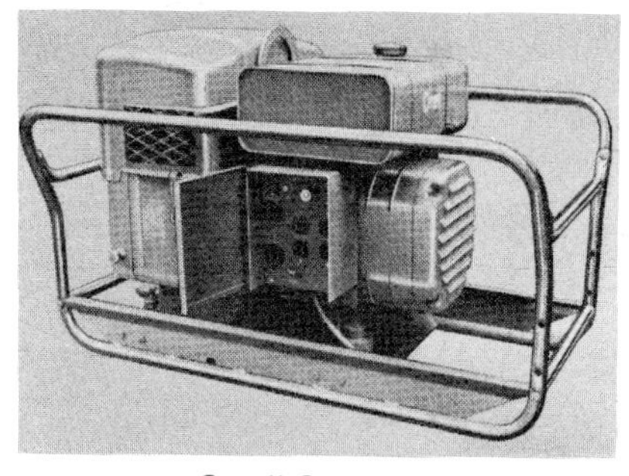

Small Generator

Onan manufactures electric generating systems for home and farm use (emergency), as well as for hundreds of other applications–industrial, commercial, etc. They offer generators ranging from a 12 volt DC, 600 watt gas powered generator, a 1,000 watt, 120 volt AC model, all the way up to a 600 kw unit with a 277/480 volt AC rating. A 5 year or 1,500 hour warranty for standby power systems (engine, generator, switch) is offered, which is something to think about if you want an alternate system.

Catalog is free

OREM'S ORGANIC GARDENS
P.O. Box 50
Petrolia, Calif. 95558

PRODUCTS:

energy publications

A publication called "Wind, Water and Sun" is the result of more than 30 years of research and development of pollution-free energy for the user. There are charts giving the energy potential of the wind, streams and the sun. There is a capacity chart that shows the amount of water a reservoir will hold and the amount of material that must be removed for construction. Research was done by putting the products to use in the field and is written from actual experience.

The book sells for $5.00

OSSBERGER–TURBINE-FABRIK
D-8832 Weissenburg in Bayern
P.O. Box 425
West Germany

Small Hydroelectric Plant

PRODUCTS:

hydroelectric plants

OSSBERGER-TURBINE FABRIK (Cont'd)

This company has more than 7,000 units in use throughout the world from Alaska to Africa that supply the electrical needs of complete towns, villages and small homesteads. Hydro-Light, their miniature power plant, occupies a space of just over a cubic yard and supplies energy for smaller premises where sufficient water is available. The unit requires minimum foundation because of the unitized construction. It only needs connections to the pressure and draft tubes and electrical gear for starting and meeting electrical needs. The regulation of this unit is handled manually. As the current consumption increases, the turbine is opened by means of a handwheel, and just the opposite for current reductions. Normal current change, such as the turning on or off of a light bulb, would require no adjustments.

The catalogs are free of charge

OWNER-BUILDER PUBLICATIONS
P.O. Box 550
Oakhurst, Calif. 93644

PRODUCTS:

methane generator plans
solar greenhouse plans

A small California-based publishing company specifically geared to low-cost, low-technology, self-help books. The three books this company has published deal with building your own solar heated pit-type greenhouse, methane generators, and just about any form of construction that would be needed for the home and homestead. The books are mainly about building construction, but have several sections devoted to alternate energies, their uses and construction.

"The Owner Built Home" is $7.50
"The Owner Built Homestead" is $5.00 for the two volume set

PINCOR PRODUCTS
Pioneer Gen-E-Motor Corp.
5841 West Dickens Ave.
Chicago, Ill. 60639

PRODUCTS:

alternators, diesel powered
alternators, gas powered
automatic load transfer panels

Pincor manufactures revolving field, self-regulated alternators that are mounted to the engine crankshaft. They range in output from 115 volts, 10.8 amps up to 115/230 volt, 34.8 amp units in the 2-pole type, and deliver, depending on the model, between 1250 and 8000 watts of 60 cycle AC current at 3600 rpm. Also available are 4-pole units with 115 volts at 13 amps up to 115/230 volt packages with 15,000 amps, powered by either gasoline or diesel air-cooled engines. The automatic load transfer panel automatically starts your stand-by power plant upon interruption of commercial power, transferring your load circuits to standby operation, and when normal power is restored, the transfer panel transfers your load back to commercial power and resets the generator for standby operation. The automatic load panel, like the alternators and other products, is of the same high quality that Pincor has been manufacturing since 1932.

Selection guide and power sheets free

PPG INDUSTRIES, INC.
One Gateway Center
Pittsburgh, Pa. 15222

PRODUCTS:

solar collectors

PPG INDUSTRIES, INC. (Cont'd)

Manufacturers of a flat plate solar collector, 34" by 76", that utilizes double-glazed tempered for durability and is backed with insulation and fitted with connectors ready for operation.

Brochures are available by writing c/o Mr. R. W. McKinley at the above address

RAYOSOL
Carretera de Cadiz 32 D
Torremolinos (Malaga)
Spain

PRODUCTS:

solar water heaters

This solar water heater consists of two parts, the so-called absorber and the tank, which are connected by tubes when they are installed. The heart of the absorber is a copper sheet, bonded either to a serpentine or to a series of vertical copper tubes. Behind the copper plate a layer of insulating material prevents the heat from escaping through the back. The whole unit is encased in a fiberglass, polyester box, with the edges and joints between the glass plates sealed by aluminum strips. The tank can be supplied in two different models to suit the space available for it. The tanks are also hot galvanized to give adequate protection against rust. Solar water heaters from this company are in constant use throughout Spain and the Mediterranean coast.

Catalog with 8 pages of color photographs and tables free

Gas Fired Boiler

RAYPAK, INC.
31111 Agoura Road
Westlake Village, Calif. 91363

PRODUCTS:

steam boilers, gas fired

These boilers are to be used with either natural gas or propane, which means they could also be used with methane. They feature 100% copper and bronze waterways and deliver between 125,000 and 1,125,000 BTU's per hour at low cost. Very little heat is lost because they use finned copped tubing instead of smooth type tubing. A thermal heat exchanger is used. Raytherms boilers are unaffected by rain, wind or debris. They utilize a unique draft system for the burner compartment housing the lifetime titanium-stainless, clog-free burners.

Illustrated product sheets free

REAL GAS & ELECTRIC CO., INC.
P.O. Box A
Guerneville, Calif. 95446

PRODUCTS:

batteries
generators, motor powered
generators, PTO powered
motorcycles, electric powered
solar collectors

REAL GAS & ELECTRIC CO., INC. (Cont'd)

generators, wind powered
inverters
wind generator manuals
wind speed indicators

The products marketed by this company consist of wind generating equipment, solar heating systems, and power back-up systems, including installation and servicing of all items sold by them. They have wind generating systems that range in size from 200 to 100,000 watts and in price from $500 to $150,000. They have an enormous amount of information on selection and design necessary to help you to make valid decisions as to selection. Several battery systems are offered from domestic and foreign sources, as well as inverters to change over to the desired voltage. In addition to wind powered generators, they also offer a complete line of generators capable of being powered by tractors, gas engines or water wheels. One of the more unusual items offered by this company is a battery powered mini-bike that has a 50 mile range at 30 mph with little or no noise, and a similar unit that attaches to a bicycle. Solar collectors suitable for hot water heating of the home or swimming pool are also offered. Instruction and schematic manuals on all Dunlite and Electro plants are available for $3.00 per set.

An extremely informative catalog is available for $3.00, and if you want to be on their mailing list as they develop new products, they are asking $1.50 to cover printing and postage costs

RELIABLE INDUSTRIES, INC.
34403 Joel St.
New Baltimore, Mich. 48047

PRODUCTS:

steam engines

Seven models of steam power plants, 4 H.P. to 200 H.P. Modern and old fashioned designs. Engines may be used for auto, marine, stationary applications. The kits are in castings form. A metal lathe, a drill press, and the necessary skill to operate power tools are required. Blueprints and plans may be purchased separately. Boilers, condensers, feed pumps and other accessories are also available.

20 page illustrated catalog-handbook $1.00

FRED RICE PRODUCTIONS, INC.
6313 Peach Ave.
Van Nuys, Calif. 91401

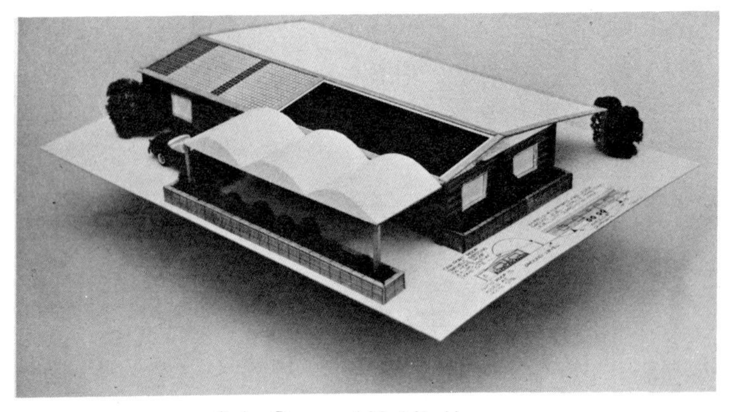

Solar Powered Mobile Home

PRODUCTS:

clocks, solar
houses, solar powered
mobile homes, solar powered
solar water heaters, cylindrical

Fred Rice Productions has designed a solar clock that can be used both indoors and outdoors with full dimensional sound response. It draws its power from the sun. They also produce the Solar/Sonic Flip-Top mobile home that draws its electrical energy from solar cells, its hot water from cylindrical solar water heaters, and its heating and cooling from the sun. This

FRED RICE PRODUCTIONS, INC. (Cont'd)

company also offers houses utilizing the same energy conversion methods. In addition, they market an efficient cylindrical solar water heater originally developed in New Zealand. This may be the only solar water heater available on the market with its storage tank integrally designed with the collector.

Portfolio of products with illustrations $1.00

W. R. ROBBINS & SON ROOFING CO.
1401 N.W. 20th St.
Miami, Fla. 33142

PRODUCTS:

solar water heaters

The solar water heaters manufactured by this company are available in two sizes, a 4′ x 12′ collector containing 120 feet of copper tubing (normally used with an 80 gallon storage tank and pump with thermostat) that sells for around $500, and a larger model, a 4′ x 14′6″ collector and 100 gallon reservoir with booster and thermostat that sells for almost $600. They have been in business for 34 years at the same address.

Illustrated brochure free

SEARS, ROEBUCK & CO.
Chicago, Ill. 60607

PRODUCTS:

air compressors, 12V
alternators, gas powered
burglar alarms, 12V
can openers, 12V
circuit breakers, 12V

Small Generator

SEARS ROEBUCK & CO. (Cont'd)

- cosmetic mirrors, 12V
- fans, 12V
- fluorescent lights, 12V
- generators, gas powered
- hand mixers, 12V
- incandescent lamps, 12V
- inverters
- refrigerators, 12V
- shavers, 12V
- toothbrushes, 12V
- water pumps, 12V
- winches, 12V

One catalog often overlooked is the Sears Catalog of Accessories for Mobile Homes, Recreational Vehicles and Camping. Many of the machines run on low voltage generators or the battery systems of campers and mobile homes, and can be used on the homestead. Refrigerators operate from 12 volts or natural gas, which automatically qualifies them for methane gas. They are available complete with freezer compartments and operate on as low as 40 watts with a 5 amp draw, which is well within the means of a small homemade system. Incandescent and fluorescent lights are also available that give better lighting than a sealed-beam light hanging from the ceiling any day.

Catalogs are available at any local Sears store in the nation

SEMPLE ENGINE CO., INC.
Box 8354
St. Louis, Mo. 63124

PRODUCTS:

steam engines

Marine steam engine kits for powering launches and houseboats. Burn either coal or wood. Engines are safe, dependable, economical, quiet, 5 to 10 HP. Engine kits are complete, not castings only. All the difficult machining work has been done. 95% of the remaining work can be done on a 10-inch lathe and drill press. Drawings are to the finest detail. Boilers and other steam power accessories available.

Information packet $1.00

SKYTHERM PROCESSES AND ENGINEERING
2424 Wilshire Blvd.
Los Angeles, Calif. 90057

PRODUCTS:

- air conditioning, solar powered
- solar house heaters
- solar stills

Only Skytherm has developed a modular roof system of interchangeable, patented units that provide thermal comfort, solar heated and night sky cooled water supply, and distilled water from solar stills. Skytherm's movable insulation panels act as a thermal valve so as to use the climate rather than overpower it. In winter, the uncovered waterjackets, which lie below the movable insulation panels, are solar heated, and the panels automatically close at night to prevent heat loss. In the summer months, heat absorbed from the room is stored in the waterjackets until the panels, which then prevent daytime solar heating, movc aside to allow night air cooling. One of Skytherm's houses was evaluated by the U.S. Department of Housing and Urban Development and was found to help meet the fossil fuel, power shortage and pollution crises, while freeing more floor space and not requiring circulation units.

No catalog is available at present but illustrated company brochures are free

SMALL HYDROELECTRIC SYSTEMS
P.O. Box 124
Custer, Wash. 95490

PRODUCTS:

hydroelectric plants

This company is the northwest sales agent for both The James Leffel Co. of Springfield, Ohio, and Ossberger-Turbinfabrik of Germany, both of which are top quality manufacturers producing highly efficient machines.

Catalog is free

SOLAR COOKERS & PARTS
2523 16th Ave. So.
Minneapolis, Minn. 55404

PRODUCTS:

solar cookers

Solar cookers are here to stay. They are simple to use and can save a fortune in fuel if people must cook. The aluminum parabolic mirrors made by this company have a reflective surface adequate and safe for both cooking and heating water. The cooker may be placed in a large sunny window or outdoors. The parabolic mirrors can also be used as molds to make more mirrors out of fiberglass, papier mache, etc.

Catalog is free

SOLAR ENERGY CO.
810 18th St. N.W.
Washington, D.C. 20006

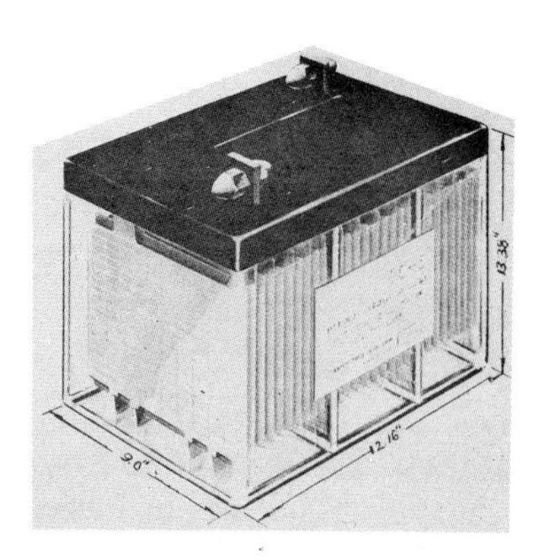

PRODUCTS:

batteries
battery chargers, solar powered
generators, wind powered
hydroelectric plants
pool heaters, solar
pumps, wind powered
solar cell banks
solar house heaters
solar stills
solar water heaters

Offered are over 60 different wind driven generators, small hydroelectric plants, solar cell banks and related equipment. "Solar Energizer," the solar cell bank marketed by this company, has a guarantee that states it will unconditionally replace any unit free of charge for the life of the owner if it fails to meet the prescribed specifications. The casing for the bank is made of an all-weather epoxy plastic with a silicon rubber protective coating. It is 3.5 x 13.75 x 0.25 inches in size and generates 12 volts at 0.167 amps. The other items carried by this company are all quality products manufactured in Europe and the U.S. Solar Energy Co. tries to be aware of all alternate energy equipment available on the world market. Also important is their System Design Manual, which will allow the reader to design a system to meet individual requirements in harmony with the local environment.

Illustrated brochure free
System Design Manual $10.00

SOLAR ENERGY DIGEST
P.O. Box 17776
San Diego, Calif. 92177

PRODUCTS:

solar publications

A monthly newsletter covering all the many facets of solar energy conversion (including wind, wave, water power and bio-conversion) and any direct or indirect conversion of energy from the elements as opposed to fossil fuel-produced energy. Want to know about solar/hydroelectric combined power system research, rocks for solar home heat bins, want to design and build your own solar swimming pool heater, etc., etc.? These and thousands of other questions are answered by William B. Edmonson in his publication collecting information from government, industrial and research sources and providing an excellent reference tool for its readers.

Subscriptions are $27.50 per year complete with photos

SOLAR ENERGY RESEARCH AND
INFORMATION CENTER
1001 Connecticut Ave., Suite 632
Washington, D.C. 20037

PRODUCTS:

solar publications

This organization publishes two bi-weekly newsletters, "Solar Energy Washington Letter" and "Solar Energy Industry Report," printed on alternate Mondays. Although still new, the

SOLAR ENERGY RESEARCH (Cont'd)

Solar Energy Center will soon become a complete resource and information outlet to serve industry, government and the private citizen, offering consulting services and a centralized information source. These publications are most competent at translating the bureaucratic jungle of energy-related legislation and can keep you clearly informed as to who's doing what, and why. They will keep you up to date on the state of the solar industry field.

"Solar Energy Washington Newsletter" subscription for one year is $75.00
"Solar Energy Industry Report" subscription for one year is $75.00
A dual subscription for both newsletters would cost $125.00

SOLAR-PAK
P.O. Box 23585
Tampa, Fla. 33622

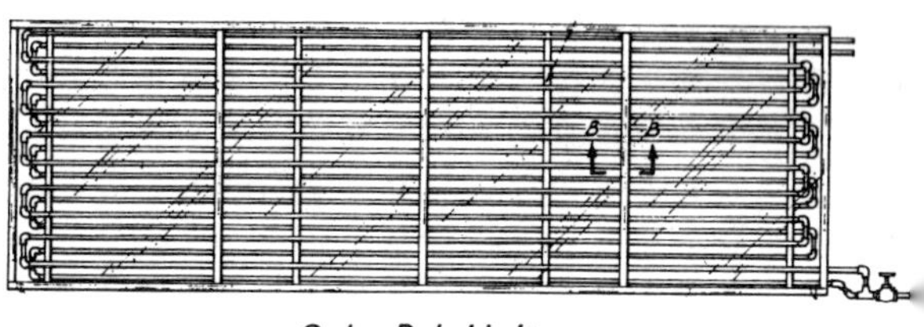

Solar-Pak Unit

PRODUCTS:

solar water heater plans

Plans prepared by this company are designed for home use as an auxiliary unit to a new or existing hot water system. The system can be modified for other uses, such as snow and ice removal for walks and driveways, or as a prime source for places with large storage capacities—car washes, restaurants, school cafeterias, etc.

Plans, material and tool list $7.00

SOLAR POWER CORP.
186 Forbes Road
Braintree, Mass 02184

PRODUCTS:

solar cell banks

Two solar cell banks are produced by this company. Module 1002 has five silicon solar cells packaged and hermetically sealed in a polycarbonate case and silicone rubber. It is capable of producing 2.2 volts and 0.6 amps in peak sunlight. The second module, SBM-6, is six model 1002 modules packaged in a weatherproof fiberglass case designed to withstand a marine environment and provides 12 volts and 0.6 amps in peak sunlight.

4 page illustrated catalog free

SOLAR SUNSTILL, INC.
15 Blueberry Ridge Road
Setauket, N.Y. 11733

PRODUCTS:

solar collector covers

The products sold by this company are coatings that are applied to glass surfaces that allow the sun to pass through when they are wet and reflect the sunlight when they are dry. They also manufacture a coating that prevents fogging from condensation inside the collector.

Catalog is sent free

SOLARSYSTEMS, INC.
P.O. Box 744
Tyler, Tex. 75701

PRODUCTS:

solar collectors

This collector is capable of producing sufficiently high water temperatures to operate a lithium bromide absorption-type air conditioner. It is also capable of providing sufficient heat for space heating and domestic hot water needs as well as many commercial, industrial and agricultural needs. The collector is marketed under the name "Solarvak," measures 4 x 8 feet, is 4 inches thick, and weighs 130 pounds dry. Solarsystems, Inc., warrants against failure to deliver the performance specified and against defective materials and workmanship for one year from the date of installation.

No catalog in print but technical sheets and test reports available free

SOLAR SYSTEMS, INC.
323 Country Club Drive
Rehoboth Beach, Del. 19971

PRODUCTS:

solar house heaters

This company makes solar heating systems for residential, commercial or industrial buildings (low temperature only, 72 to 85 degrees) compatible with existing roofs, regardless of direction and pitch, and compatible with existing duct (hot air) or radiator (hot water) heating systems. The present estimated complete system cost, installed in a home, is between $2.25 and $2.50 per square foot of floor space, depending on insulation, windows, shade, prevailing winds, etc. They claim their system will provide 100% of a building's heating requirements 80% of the time or better, depending on location, and, at present fuel prices, should pay for itself in fuel savings within 8 to 10 years, unless fuel prices double, in which case it might only take 4 to 5 years. The system is designed for a 20-year or longer life span, with a one year guarantee against defects in material and workmanship. Owners could install the system themselves, saving on-site labor, but most systems would be installed by existing plumbing, heating and air conditioning contractors.

Brochure and information $1.00

SOLAR WATER HEATER CO.
P.O. Box 1872
Coral Gables, Fla. 33134

PRODUCTS:

- pool heaters, solar
- solar water heater plans
- solar water heaters

An experienced, reliable company, Solar Water Heater Co. has been in business since 1923 building and installing thousands of efficient, long lasting systems. Their system is constructed and installed to get maximum benefit from the sun, and is suitable to install on any type of building. A pool heater is also available that uses the same style collector as the smaller units but with a larger capacity. This company also sells plans to construct either of these systems complete with storage and pumping networks.

Illustrated catalog &7.50 postpaid and plans are $45.00 postpaid

SOLAR WIND
P.O. Box 7
East Holden, Me. 04429

PRODUCTS:

batteries
generators, wind powered
inverters
towers
wind publications
wind speed indicators

Wind-Powered Generator

Wind generators producing from 12 volts D.C. up to 230 volts A.C. are available through this company, which imports them from Germany, Switzerland and Australia. They also handle American-made generators. Individual components may also be purchased separately, such as Rohn guyed towers and Dunlite towers, D.C. to A.C. inverters, storage batteries, and wind speed indicators. Solar Wind offers complete generating systems with the windplant, controls, tower, and battery sets, ready for installation on the proposed site. They also publish a 29 page pamphlet that will allow the reader to compute his electrical requirements and to design a wind generating system to fit his personal needs.

Illustrated pamphlet and supplementary brochures cost $2.00

SOL-THERM CORPORATION

7 West 14th St.
New York, N.Y. 10011

PRODUCTS:

solar water heaters

Manufacturer of a flat plate solar water heater which can be installed as an add-on or "piggy back" to an existing electric, gas or oil hot water tank. This company has fifteen years' experience with solar water heating and offers commercial and industrial applications in addition to residential applications. They also offer engineering assistance in solar water heating projects.

For brochure send self-addressed, stamped business envelope

SPECTROLAB

12484 Gladstone Ave.
Sylmar, Calif. 91342

PRODUCTS:

solar cell banks

Spectrolab has developed a system of stacking solar cell modules so that they can supply any current at any voltage desired. The solar cells are mounted in rows on boards and stacked between two pieces of channeled aluminum with

SPECTROLAB (Cont'd)

adjustable legs so that they can be set at any angle desired to use the sun's energy for maximum efficiency. Solar power systems must be designed specifically to meet the current and voltage requirements of the load at the installation site. Most of Spectrolab's systems are installed in communication systems such as remote VHF/UHF repeaters, microwave relays, environmental and pollution monitors, etc., so they show evidence of being highly reliable and offer quiet and clean electrical energy.

15 page booklet and information sheets free

SUNDIALS & MORE
P.O. Box H
Playground Road
New Ipswich, N.H. 03071

PRODUCTS:

sundials

This company manufactures, imports and markets sundials that are primarily sand-cast in bronze. If you are going to heat your house or your water with the sun, you may as well take full advantage of the energy and use it to tell the time of day.

Catalog is free

SUNSOURCE
9606 Santa Monica Blvd.
Beverly Hills, Calif. 90210

PRODUCTS:

solar water heaters

The Sunsource solar water heater is the highly efficient collector imported from Miromit in Israel. The key to this particular system's suitability in the U.S. is the unique "selective black" coating developed by Dr. Tarbor in Jerusalem. The coating is more efficient than a black painted absorber of the same size because of better heat transfer qualities. In experimental installations these panels have been used to drive a vapor cycle turbine engine which was used to drive a generator. This company will probably be producing these in the U.S. under license before June of 1975.

There is no charge for the catalogs

SUNWATER COMPANY
1112 Pioneer Way
El Cajon, Calif. 92020

PRODUCTS:

- pool heaters, solar
- solar stills
- solar water heaters

Sunwater's solar pool blanket is made of durable plastic materials especially compounded to withstand the sun's ultraviolet rediation and chlorinated swimming pool water. The covers trap the sun's rays to the entire depth of the pool and eliminate evaporation and heat loss. The blanket insulates with sealed air pockets that cover its surface and act as magnifying glasses to heat the water. Their solar still, which purifies water, is a shallow pan with a sloping glass cover. Water in the pan is heated by the sun's rays, pure water evaporates, rises, condenses on the underside of the glass, and runs down the collection trough. The pans and reservoirs are guaranteed for three years, but many have been in use for much longer, so there is no inherent limit to how long they will last. The solar water heater is designed to be simply attached to an existing system at full city pressure, without necessity of a separate or enlarged storage reservoir. A pressure relief valve is supplied in case excessive temperature produces steam pressure. Any plumber or handyman can make the installation as no special skills or tools are required.

Product information sheets free

TECUMSEH PRODUCTS CO.,
ENGINES DIVISION
900 North Street
Grafton, Wisc. 53024

PRODUCTS:

engines, gas powered
gear drives

In 1884 John Lauson, with his uncle George Lauson and J. H. Openberg, opened a machine shop using a windmill for power. It burned down in the following year. Shortly after this they started manufacturing traction steam engines and continued until 1895, when they began making gasoline engines. They now manufacture small gasoline engines ranging from 2 hp to 16 hp, with just about any horsepower you could want in between. The engines are designed for lawnmowers and industrial usage. Tecumseh warranties their engines for a period of one year. If anything goes wrong during that time they will repair or replace it at no charge to the customer. They also manufacture right angle gear drives, T-drives and transmission units which could be used in conjunction with gas engines or water wheels to give you a selection of speeds and torques needed to do work that is now impossible.

Catalogs that are full of technical information and illustrations are free

THOMASON SOLAR HOMES, INC.
6802 Walker Mill Road S.E.
Washington, D.C. 20027

Solar Heated House

PRODUCTS:

- air conditioning, solar powered
- solar house heaters
- solar workshops

This company has been one of the leaders in solar energy development for residential usage since 1959. Dr. Harry Thomason invented and patented the world's first really successful solar heating system using his own residence in Washington, D.C. During his first winter, in 1959, the total fuel bill for his 3-bedroom solar heated home was only $4.65. Encouraged by that success with solar heat collectors and heat storage apparatus, the inventor expanded his research and development to include air conditioning of high quality to save energy and money for the homeowner. Courses and workshops are held at George Washington University, Edmund Scientific Company, and at the company's offices. They are designed to help others learn what solar energy can do to save money, to save our precious energy supplies, and to reduce pollution.

Printed sheets with illustrations and photos are free
Workshop courses lasting five days are $425.00

TRANTER, INC.,
PLATECOIL DIVISION
735 Hazel St.
Lansing, Mich. 48909

PRODUCTS:

- solar collectors

TRANTER, INC. (Cont'd)

Tranter manufactures a highly efficient flat plate solar collector. The panels, called "Econocoil," are very high quality and are easy to handle.

Informative catalog $1.00

UNIVERSITY OF FLORIDA
SOLAR ENERGY & CONVERSION LAB
Gainesville, Fla. 32611

PRODUCTS:

solar publications

Adding to the growing level of development in solar hardware, Dr. E. A. Farber has done an impressive amount of work on such items as solar stills and heat engines. You can get copies of papers describing much research.

Apparently information is free

VISUAL PURPLE
P.O. Box 979
Berkeley, Calif. 94701

PRODUCTS:

publications

"The Natural Energy Workbook" is meant as a guide to natural energy utilization. It is also a manual and a how-to-build reference for several systems using wind, water, sunlight and methane gas for power. Information includes graphs not available elsewhere, including tables for calculating optimum collector size and tables for deriving optimum wind plant size.

The book sells for $3.95

VITA (VOLUNTEERS IN TECHNICAL ASSISTANCE), INC.
3706 Rhode Island Ave.
Mt. Rainier, Md. 20822

PRODUCTS:

- hydraulic ram plans
- solar water heater plans
- windmill plans

VITA is a private non-profit association of 6,000 volunteers—architects, educators, agronomists, accountants, day care specialists, chemists, pharmacists, etc.—who supply technical assistance to persons or groups in this country and abroad who cannot obtain the help they need from local sources. The service is free and the primary emphasis is on rural development efforts. They have responded to more than 20,000 such requests in the past 15 years. The transfer of technology from those who possess special knowledge and experience to those who do not is the main purpose of the group.

List of publications is free
"Village Technology Handbook" is $7.00
"Hydraulic Ram for Village Use" is 25¢
"Low Cost Windmill for Developing Nations" is $1.00

THE WHOLE MOTHER EARTH WATER WORKS
Green Spring, W. Va. 26722

PRODUCTS:

- hydraulic ram water pumps
- hydraulic ram water pump kits
- hydraulic ram water pump plans

Edward Barberie is one of the few modern day benefactors of the poor homesteader who is without a means to move water in a direction other than its own. He will give a hydraulic ram to anyone who has a need of it and who can't afford to buy one. The hydraulic ram is made from his own design, and is made of standard, though not always easy to get, plumbing parts that retail for around $25.00. If you have the $25.00 for the parts, he has a set of plans that allow you to build your own following his easy to read and clear directions. If you run into any problems, he will try to put you in touch with someone who lives in your area, or try and do what he can from Green Spring, West Virginia. Included with the plans are tables that let you choose the right drivepipe size and ram to use for the flow, and the fall from the source to the ram. Hard to find items such as drivepipes and other components of the ram can be obtained through this company at low cost, or they will give you the address where to get them. The free hydraulic ram program is paid for by selling merchandise and taking 10% of the money to help offset the expense.

Catalog and plans are free, but stamps to cover the mailing probably wouldn't hurt

WINCO
2201 E. 7th St.
Sioux City, Iowa 51106

PRODUCTS:

- generators
- generators, motor powered
- generators, PTO powered
- generators, wind powered

Two-Bearing Generator

The Winco company is one of the oldest and most prestigious companies manufacturing and selling generating equipment in the United States today. In addition to its other power generating equipment, Winco makes four types of two-bearing generators that can use any type of drive source that you have. They range in output from 1,000 to 15,000 watts, and have receptacles built in to the cases. This company's generators are designed for either direction of rotation, standard being counter-clockwise facing shaft extension, and have both 2- and 4-pole revolving armatures inherently regulated, self-excited, drip-proof, and two ball bearing support. The 2-pole generators operate at 3600 r.p.m. and, depending on model, put out 115 volts at 8.7 amps for the 1,000 watt model up to 115/230 volts at 65 amps for the 15,000 watt unit.

"Wincharger," Winco's wind generator, delivers 12 volts, 200 watts in a 7 m.p.h. wind. This generator has an all-around performance that can't be equalled in the 12 volt lighting field because it starts charging in a 7 m.p.h. breeze, and has an output of 15 amperes in a wind velocity of 23 miles per hour. Its 6 foot propeller is built of the famous Albers Air-Foil principle, with machine-made, perfectly balanced, copper armored leading edge and weather-proof varnished blades. The blades are connected to the generator via an air-brake governor that operates by cenrtifugal force so that when the wind exceeds 23 m.p.h. the governor automatically opens and spreads away from the propeller, slowing it down using air resistance. The governor also acts as a flywheel to maintain even propeller speed and eliminate vibration in gusty winds.

Portable electric generators are also available from 1,000 to 12,500 watts powered by gas and diesel fueled engines. All are mounted on skid type platforms and are complete and ready to go, with both manual and electric starters. They feature automatic idling controls, so that when no current is being used the engines will idle, and when there is a load on the circuit the engine speeds up the generator to match the requirements. Receptacles are mounted on the control panels.

All Winco's generators are AC except for the 12 volt

Wincharger unit, and use extra windings in the fields so that extra power is provided when there is a heavy starting load on the system.

Illustrated brochures free

WINDWORKS
Box 329, Route 3
Mukwonago, Wisc. 53149

PRODUCTS:

generators, wind powered
wind energy, consultation
wind energy workshops
wind generator parts
wind publications

Windworks is a small research and design group that began under the direction and sponsorship of Buckminster Fuller. They began with the development of a paper honeycomb technique of blade construction that allows individuals to construct low-cost, efficient airfoils. Their present products include two sets of plans: a 12 foot diameter, high speed, solid foil wind generator, and a 25 foot sail wind generator. For hard to get items and those that are expensive through retail outlets, Windworks offers components for the 12 foot wind generator such as honeycomb blocks for blade construction, 12 volt 85 amp Motorola alternators, or blade shafts. Consultation (aerodynamic, structural engineering, feasibility studies, etc.) services run about $20.00 per hour plus expenses, depending on the job. A wind energy bibliography is available that lists relevant books, trade publications and catalogs of manufacturers. Wind energy workshops, usually taking the form of slide-show seminars, are held for schools or organizations on request, and last from one to two days.

Illustrated catalog sent for donations
"Wind Energy Bibliography" $2.00

ZOMEWORKS CORP.
P.O. Box 712
Albuquerque, N.M. 87103

PRODUCTS:

- solar energy slide sets
- solar house heater plans
- solar house heaters
- solar powered ventilators
- solar publications
- solar tracker
- solar water heater plans
- solar water heaters

Zomeworks was organized in the spring of 1969 by a group who had been working together for several years building zomes and solar heating systems. This company also designs and constructs solar heating systems for both water and space heating. They favor designs that work by themselves, such as the Skylid and the convective rock storage systems, or those that can be operated manually, such as the doors for their drumwall. The advantage of not relying on electric pumps and fans is that the systems are simpler to maintain and are suitable for remote locations.

In the Drumwall structure, south facing walls are made of 55 gallon drums filled with water in racks behind glass, with the outside facing ends painted black. The winter sun warms the water in the barrels through the glass, and the heat radiates into the rooms. The large insulating doors are lowered in the mornings in the winter and allow the sun to shine in, and the barrels act as reflectors to intensify heat from the sun. The doors are raised when the sun goes down to prevent heat loss. During the summer the doors are kept closed and the cool water in the drums acts as a sink for the heat in the air, keeping the room at a comfortable temperature. The drumwall is a specially satisfactory method of heating and cooling as the drums are an integral part of the structure of the house.

A bin of rocks called the Air Loop Convective System, usually beneath or in front of the house, is heated by air flowing by convection through a collector. When heat is needed in the house, a vent is opened and the warm air rises into the room. In the summer the rock bin can be cooled at night and cool air vented into the house during the day.

Skylids are insulated louvers that are placed inside a building or beneath skylights, glass roofs or vertical windows. They open during sunny weather and close by themselves during very cloudy periods and at night. When the skylids are closed, they become an effective thermal barrier, greatly reducing heat losses through glazed openings. This allows one to have large glass areas that let solar heat and light into the building during the day without having large heat losses through these glass areas at night. These louvres have the additional feature of allowing one to regulate, by a manual override, the amounts of heat and light that a skylight admits into a room.

A collection of articles written by Steve Baer, the originator of the zome style construction, originally published in bi-weekly installments in the "Tribal Messenger," are put together in a book called the "Solar Booklet." This covers solar heat, heat exchangers, convection and other related items. A solar slide set consisting of 21 slides showing different types of solar heating installations, including solar heated greenhouses, drumwalls, skylids and other solar heating devices, is available for $12.00, including postage and mailer.

Illustrated catalog and information sheets free
"Solar Booklet" $3.00 postpaid

MASTER INDEX

CAN opener, 12 volt
Sears Roebuck and Co.
CHAINS and sprokets
Culliman Wheel Co.
CIGARETTE lighters, solar, (see Solar cigarette lighters)
CIRCUIT breakers, 12 volt
Sears, Roebuck and Co.
CLOCKS, solar
Fred Rice Productions, Inc.
COLLECTOR covers, solar, (see Solar collectors covers)
COLLECTORS, solar, (see Solar collectors)
COMPRESSORS, air, 12 volt, (see Air compressors, 12 volt)
CONSULTATION wind energy, (see Wind energy, consultation)
COOKER, solar, (see Solar cooker)
COSMETIC mirrors, 12 volt
Sears, Roebuck and Co.
CYLINDRICAL solar water heaters, (see Solar water heaters, cylindrical)
DIESEL powered alternators, (see Alternators, diesel powered)
ELECTRIC powered motorcycles, (see Motorcycles, electric powered)
ELECTRIC valves, 24 volt
Airborne Sales Co., Inc.
ENERGY course, solar, (see Solar energy course)
ENERGY publications
Orems Organic Gardens
ENGINES, gas powered
Kohler Co.
Tecumseh Products Co.
ENGINES, steam, (see Steam engines)
FANS, 12 volt
Airborne Sales Co., Inc.
Sears, Roebuck and Co.
FANS, 24 volt
Airborne Sales Co., Inc.
FLOURESCENT lights, 12 volt
Sears, Roebuck and Co.
Theodore Bargman Co.
FLUID pumps, 12 volt
Aircraft Components, Inc.
FREQUENCY changers
Nova Electric Mfg. Co.
FRESNEL lenses
Cryton Optics, Inc.
FURNACE, solar, (see Solar furnace)
GAS fired steam boilers, (see Steam boilers, gas fired)
GAS powered alternators, (see Alternators, gas powered)
GAS powered engines, (see Engines gas powered)
GAS powered generators. (see Gene rators, gas powered)
GEAR drives
Tecumseh Products Co.
GEAR trains & motor
Airborne Sales Co., Inc.
GENERATORS, all types
Airborne Sales Co., Inc.
Fesco
Winco
GENERATORS, gas powered
Kohler Co.
Onan Division-Onan Corp.
Sears, Roebuck and Co.
GENERATOR manuals, wind, (see Wind generator manuals)
GENERATORS, motor powered
Fesco
Real Gas & Electric Co., Inc.
Winco
GENERATOR parts, wind, (see Wind generator parts)
GENERATOR parts, hydrogen, (see Hydrogen generator parts)
GENERATOR plans, methane, (see Methane generator plans)
GENERATOR plans, steam, (see Steam generator plans)
GENERATOR plans, wind, (see Wind generator plans)
GENERATORS, AC-DC, (see AC-DC generators)
GENERATORS, steam, (see Steam generators)
GENERATORS, PTO powered
Onan Division-Onan Corp.
Real Gas & Electric Co., Inc.
Winco
GENERATORS, wind powered
Automatic Power Division
Budgen & Associates
Edmund Scientific Co.
Elektro GMBH
Environmental Energies, Inc.
Fesco
Negeye Laboratories
Real Gas & Electric Co., Inc.
Solar Energy Co.

Presents

the

Finder's Guide Series

FINDER'S GUIDE No. 1

Joseph Rosenbloom

This book offers the do-it-yourselfer a complete directory of companies and equipment available for many diversified projects and plans. This indexed directory solves the problem of finding out "who" makes "what." There is something here for every taste and level of skill.

288 pp

LC 73-92459 $3.95

Kits and Plans for the Budget Minded

CRAFT SUPPLIES SUPERMARKET

FINDER'S GUIDE No. 2

Joseph Rosenbloom

A well illustrated and indexed directory of craft supplies. Thousands of products, including materials, kits, tools, etc., from over 450 companies, are analyzed from their catalogs.

224 pp, ill., August, 1974
LC 74-84298 $3.95

THE COMPLETE KITCHEN

FINDER'S GUIDE No. 3

Anne Heck

This book is a comprehensive guide to hard-to-find utensils, and describes the companies supplying such utensils as well as giving information about their catalogs. Many illustrations of unusual or interesting utensils.

96 pp, ill., September, 1974
LC 74-84299 $2.95

SPICES, CONDIMENTS, TEAS, COFFEES, AND OTHER DELICACIES

FINDER'S GUIDE No. 6

Roland Robertson

Answers difficult questions involved with finding and purchasing unusual ingredients, beverages and foods which are difficult to obtain locally. This illustrated and indexed directory is highly browsable, to say nothing of gastronomically stimulating.

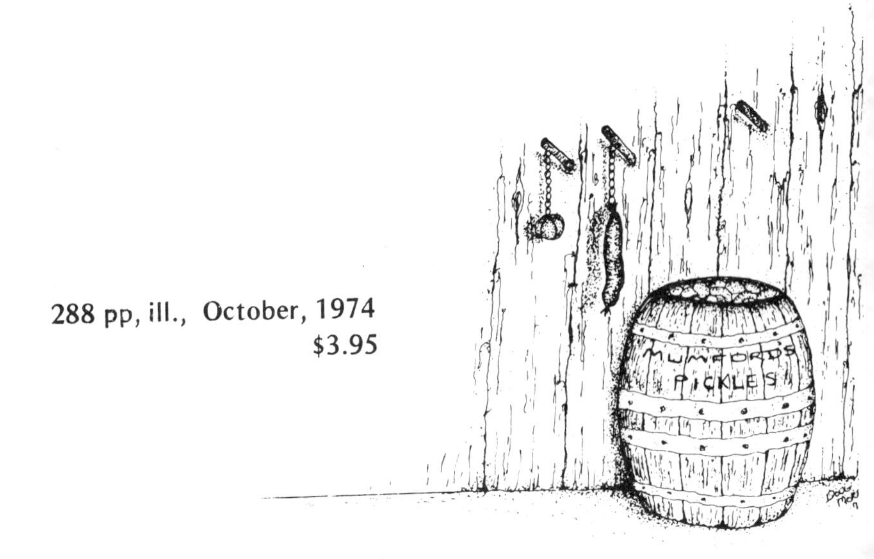

288 pp, ill., October, 1974
$3.95

COUNTRY TOOLS

Essential Hardware and Livery

FINDER'S GUIDE No. 7

Fred Davis

Locates sources for the otherwise difficult to find tools essential to country living. This book covers everything from bell scrapers through goat harnesses to spoke shavers. An indispensible guide for the country resident working his land.

272 pp, ill., October, 1974
$3.95

THE Scribner Library

America's Quality Paperback Series

CHARLES SCRIBNER'S SONS

Shipping and Billing Departments
Vreeland Ave., Totowa, New Jersey 07512

Order Blank

Dear Sirs:

I believe your new series "FINDER'S GUIDES" fills a definite need for information and I would like to order:

QUANTITY	TITLE	TOTAL
	copies of KITS AND PLANS @ $3.95 ea.	
	copies of CRAFT SUPPLIES SUPERMARKET @ $3.95 ea.	
	copies of THE COMPLETE KITCHEN @ $2.95 ea.	
	copies of HOMEGROWN ENERGY @ $2.95 ea.	
	copies of SPICES, CONDIMENTS, TEAS, COFFEES, AND OTHER DELICACIES @ $3.95 ea.	
	copies of COUNTRY TOOLS @ $3.95 ea.	
	copies of ALL OF THE ABOVE BOOKS ($21.70 Total)	

NOTES

NOTES

NOTES

NOTES